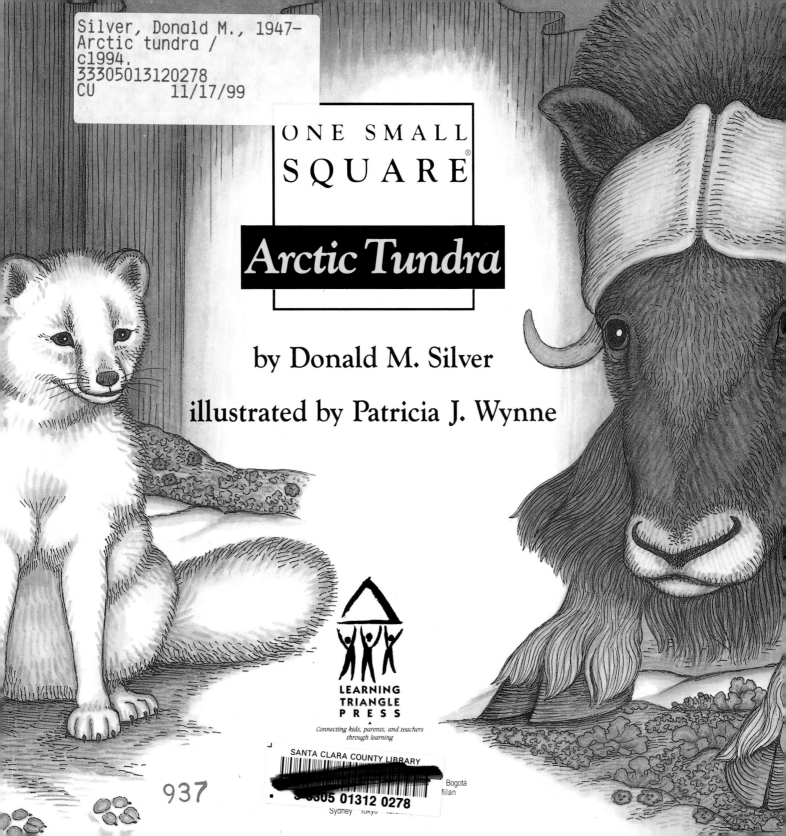

ONE SMALL SQUARE®

Arctic Tundra

by Donald M. Silver

illustrated by Patricia J. Wynne

LEARNING
TRIANGLE
PRESS
Connecting kids, parents, and teachers
through learning

Bogotá
Milan
Sydney   Tokyo

Every plant and animal pictured in this book can be found with its name on pages 40–43. If you come to a word you don't know or can't pronounce, look for it on pages 44–47. The small diagram of a square on some pages shows the distance above or below the ground for that section of the book.

For Captain A.M. Wynne,

my father, who taught by example — P.J.W.

Our sincere thanks to Maceo Mitchell for his unsung, but appreciated, efforts; Thomas L. Cathey for his continued support of this series; Dr. Robert Rockwell of the American Museum of Natural History for sharing his insights into how the tundra works; Karen Malkus for maintaining her focus and enthusiasm for these books; and Jan Rosenbaum for his clever activity suggestions.

Text copyright © 1994 Donald M. Silver.
Illustrations copyright © 1994 Patricia J. Wynne.
All rights reserved.
One Small Square® is the registered trademark of Donald M. Silver and Patricia J. Wynne.

Printed in the United States of America. Except as permitted under the United States Copyright Act of 1976, no part of this publication may be reproduced or distributed in any form or by any means, or stored in a database or retrieval system, without the prior written permission of the publisher.

Library of Congress Cataloging Number 97-074101
ISBN 0-07-057927-X
1 2 3 4 5 6 7 8 9 QPD/QPD 9 0 2 1 0 9 8 7

Whether you are outside or at home, always obey safety rules! Neither the publisher nor the author shall be liable for any damage that may be caused or any injury sustained as a result of doing any of the activities in this book.

Introduction

Here are some riddles. Where does it rain or snow very little, yet the ground is always full of water? Where does one winter night last for weeks or months? Where does the sun never set during part of the summer? Where do strange colorful curtains of light hang in the sky, attached to nothing? Where is there ice in the ground on even the warmest days? Where might you not see a living thing, though millions of animals flock there every year? What place has a name that means "treeless," although trees grow there?

The answer to each of these riddles is the same: the tundra. The tundra is the land of bears and hares and howling wolves. It is where a fox can dive into the snow and capture a lemming it cannot see. On the tundra a

3

hundred-year-old tree may be less than one foot tall. And flowers turn to follow the sun as it moves across the sky.

Nearly everything about the tundra seems mysterious or puzzling at first. Why is the winter there so long and so cold? How can some tundra animals survive being frozen alive? What makes birds travel thousands of miles just to reach the tundra?

The best way to figure out how the tundra works is to explore it. But that isn't easy. It may be far away from where you live. But even if you live close by, for much of the year the tundra's weather is dangerously cold.

If you can't go to the tundra, this book will bring one small square of tundra to you. There are activities you can do wherever you live to help you solve some of the tundra's mysteries and puzzles. And you can do almost all of the activities in your home or yard or in a park, using tools such as the ones on this page.

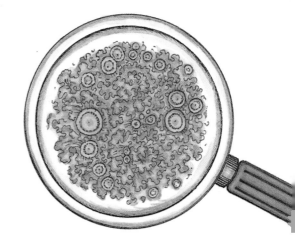

When you look through an inexpensive magnifying glass at plants and animals, you will discover details about how living things are adapted—fitted—to live where they do.

Get out your boots to walk in snow or mud. Wear gloves on icy winter days. Note in your book how a flashlight and a thermometer can help solve tundra riddles.

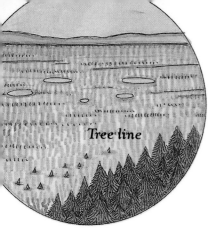

Tree line

The last of the forest trees form the tree line in the far north. Beyond this line lies the tundra. There, no full-sized trees are able to grow.

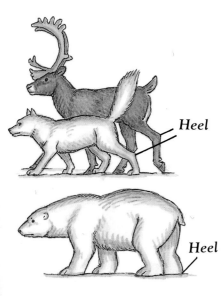

Heel

Heel

You walk on your feet. So does a bear. But a caribou walks on its toes, as does a wolf. What about a dog or a cat?

One Small Square of Arctic Tundra

Imagine you are in an airplane flying north. There are no clouds to block your view. You look down on cities, towns, farms, rivers, roads, perhaps mountains.

After a few hours all you see below are trees. A vast forest carpets the ground. At the forest edge, there are fewer and fewer trees, until they seem to disappear. Now the land is mostly flat. If your plane doesn't touch down or turn, you will soon be over the Arctic Ocean, at the top of the world.

Did you miss the tundra? Not at all. You are gazing at it. The tundra is the mostly flat land between the end of the forest and the ocean.

This book explores only one small square of tundra. The square, shown here, is 6 feet long and 6 feet wide—about the size of a camping tent. This small square of tundra is in the area called the Arctic, one of the coldest parts of Earth.

Come discover the Arctic tundra, first in winter, then in summer (page 20). You will find out what it takes to stay alive above and below the snow. When the snow melts, watch what happens to the small square. Keep an eye out for a polar bear strolling by; a fox changing color; and caribou, which sure look a lot like reindeer.

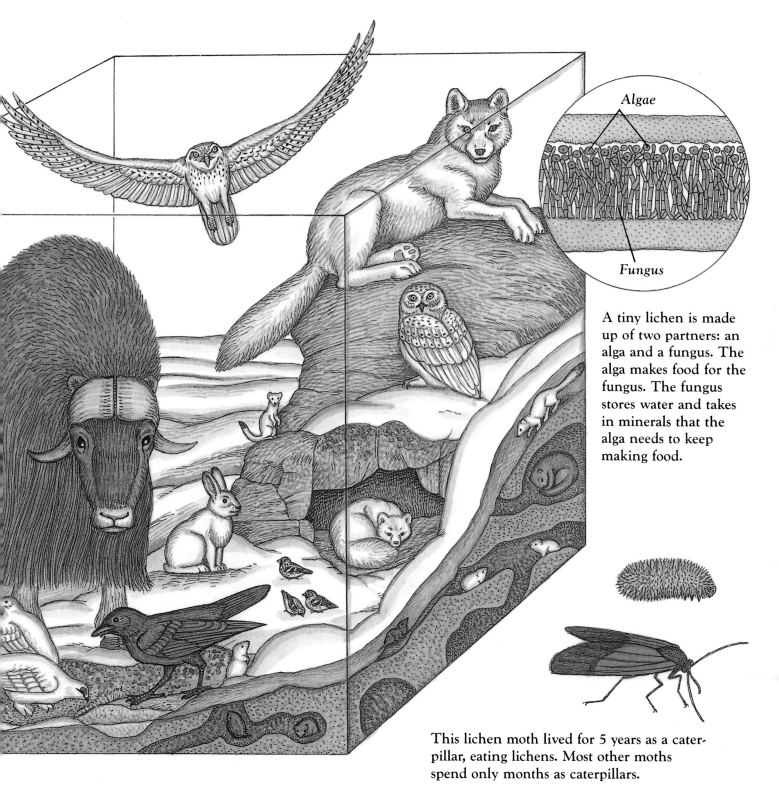

Algae

Fungus

A tiny lichen is made up of two partners: an alga and a fungus. The alga makes food for the fungus. The fungus stores water and takes in minerals that the alga needs to keep making food.

This lichen moth lived for 5 years as a caterpillar, eating lichens. Most other moths spend only months as caterpillars.

Small tundra animals, beware! The gyrfalcon is hunting for a meal. Do whatever you can to avoid this predator's hooked claws and beak.

Winter Already?

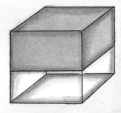

It is the end of September. There is snow on the ground. The air is always cold.

During the day the sun never rises very high in the sky. Nor does it warm the small square enough to melt even the few inches of fallen snow. Autumn may be starting where you live, but on the tundra it is already winter!

On most days the small square seems deserted. Today though, the wind blew away some of the snow. Snowshoe hares and ptarmigan stopped by to feed on twigs, buds, and berries that had been hidden by the snow.

What luck for ravens. The wind blew the snow off the remains of a dead musk ox. All the hungry birds have to do is feast.

Ptarmigan foot

With "snowshoe" feet and white feathers for winter camouflage, ptarmigan blend into the snow.

The next stop for these Arctic redpolls is south, at the edge of the tundra.

The hares hardly feel the chill through their thick fur. The ptarmigan depend on their feathers to keep body heat in and cold out. Before the first flakes, the hares grew thicker fur, the birds more feathers—even on their feet. These added coverings turned their feet into "snowshoes" that don't sink into the snow.

Hares and ptarmigan aren't the only hungry animals around. Falcons and foxes hunt them for food. These predators must have sharp eyes, for white fur and white feathers almost disappear against white snow.

Snowshoe hares, as well as the other animals shown here, make their own body heat. They grow fur (or feathers) to stop heat loss to the cold air. But they must eat to get enough energy to keep making heat.

Your Tundra Notebook

Use words and pictures to record in a notebook the activities you do. Add to your notebook by pasting in tundra news and photographs.

That Sinking Feeling

After a snowfall or heavy rain, put on boots and go for a walk in the backyard or park. How far do you sink in the snow or mud? Look for tracks left by birds and other animals. How far did they sink? Do any tracks look like those below—from the small square? If you know someone with snowshoes, ask if you may use them. Now what happens when you walk on the snow?

snowshoe hare track

wolf track

bear track

ermine track

The grizzly bear doesn't have to put up with the tundra winter. It sleeps through the cold months in its den. The bear stays warm by using its body fat for energy to keep making heat.

Down feather

Outer feather

Have you seen birds fluff their feathers on chilly days? Tundra birds do this too. When fluffed, outer feathers and inner downy feathers trap air. Escaping body heat warms this air, which stays in a snug layer around the bird.

Back Off!

It has snowed a few times—never a blizzard, just a little here and there. Now 6 inches of snow covers the ground. It's hard to tell where the small square begins or ends.

The musk oxen couldn't care less. The square is only one stop on their long walk across the tundra. What these big, shaggy animals do care about is finding food and staying far away from hungry wolves. Wherever the musk oxen are, they paw through the snow, searching for twigs and other plant parts to nibble.

The November sun hardly rises. It stays low in the sky and sets again about five hours later. During this short day the temperature may not reach a high of even 0°F (−18°C).

The musk oxen are tired. They plop down on the square for the night. But soon one of them jumps to its feet, waking the others. The musk oxen sense wolves nearby. Quickly they form a circle facing outward. They lower their heads to show off their sharp, curved horns. The message is clear—back off or else! The wolves turn away. They won't attack, at least not tonight.

No matter how cold it gets, day or night, the musk oxen keep toasty warm. Their long, dark fur coats take the bite out of the attacker that won't back off—the tundra winter.

Young musk oxen would rather be running, jumping, or butting heads. But they know when to turn to the herd's bulls and cows for safety.

Woolly fur (qiviut)

Guard hairs

A musk ox's fur coat has two layers. The woolly inner layer keeps in body heat. The outer layer of long guard hairs keeps out wind and water.

Given enough warning, musk oxen will panic and run from wolves.

Out of Sight

Aurora borealis (northern lights)

The icy wind howls. It *is* the wind, isn't it? Or is it a pack of wolves breaking the eerie silence?

The moon lights the snow. It *is* night, isn't it? Maybe not! For more than a week the sun has not appeared. Without the sun, it has been dark 24 hours a day.

One thing is certain: At −50°F (−45°C) it is bitterly cold. What brings animals out in the frozen dead of winter? Not ringing in the new year, but hunger. That's why the snowy owl is swooping down on a lemming that is in the wrong place at the wrong time. It's why an Arctic fox is digging out a half-eaten vole it hid in the snow. And why the wolf pack crossing the small square is hunt-

Fur on the bottom of the Arctic fox's paws keeps out frost as the fox digs up food.

The long, rounded wings and soft feathers of the snowy owl make hardly a sound as the large hunter drops down to grasp its favorite food —a little lemming.

Foxes and wolves fluff their fur to trap air and stay warm.

ing for anything it can find, dead or alive.

Soon the sun will start to shine again on the small square. Until then, at least the hunters have the moon to see by. But the moon doesn't give off its own light. Moonlight is light from the sun that bounces off the moon. And what about those colored lights (aurora borealis) hanging in the sky? They "turn on" when particles streaming from the sun reach Earth and cause gases in the air to glow like neon signs. The sun may be out of sight, but it still manages to deliver light.

Watch that wolf with the tail raised high. He is the leader of the pack. When he runs, the pack runs. When he rests, so do the other wolves. When he lunges at prey, the pack attacks. The only other wolf with that much power is his mate. When she leads, the pack does what she wants too.

Tilt

Take an orange and push a pencil or a stick through it from top to bottom. Your orange now is Earth, with the North Pole at the top. Draw a circle about ½ inch from the top, as shown. The small square is just above this line.

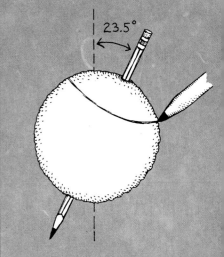

23.5°

Tilt the orange as shown. That's how Earth is tilted. Ask an adult to remove the shade from a lamp. The light bulb is the sun. Walk slowly around the lamp without changing the tilt of the orange. Rotate the pencil as you go. This is like the Earth's spin—making night and day. Note how much light shines above the line (your "Arctic Circle") as you walk.

When the top half of the Earth is tilted away from the sun, there is no day, just night, in the Arctic tundra. As Earth moves around the sun, day returns to more and more tundra. But for half the year, the North Pole has night without day.

Where the Action Is

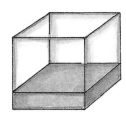

The big chill refuses to go away. With each passing week, fewer and fewer animals visit the small square. Then one morning an Arctic fox on the prowl stops by and stares at the snow. Something the fox cannot see is moving below, but the predator can hear and smell prey.

Hidden beneath the snow is a world of tunnels alive with the lemmings and voles that dig them. These cousins of mice and hamsters cannot tunnel into the ground because it is frozen solid all winter. No matter. Their snow tunnels are ideal hideaways.

Throughout the winter, lemmings and voles scurry

The only thing on the Arctic fox's mind is pouncing on a lemming. If the fox can't kill enough food, it follows a polar bear or wolves and settles for leftovers.

Snow hides and protects busy lemmings and voles. But it also blankets plants and animals waiting for warmer days to start moving and growing. Some are insects frozen in soil, but still alive.

Born under the snow, baby lemmings drink their mother's milk to survive.

about in search of buried plants to nibble. For air flow, they dig chimneys to the surface. Like a blanket, the snow cover keeps frosty winds from harming not only the animals but also the plants and soil. It may be −30°F (−35°C) where the fox is, but it is a "comfortable" 20°F (−7°C) for the furry lemmings. The lemmings are so snug in their tunnel maze that they even have babies in winter.

Of course, no hide-out is totally safe from predators. A long, slim ermine may squeeze down a chimney and through a tunnel. Or, at just the right moment, an Arctic fox may dive into the snow and catch a lemming before it can scurry away to safety.

Frozen caterpillar

The ground squirrel hibernates. All winter it sleeps, its heart and breathing slowed and its temperature lowered. Stored body fat keeps it alive until spring.

No Wonder

The sun is low in the sky early in the morning and at dusk. The sun is highest at midday. Go out and stand in the sun at these times. When do you think more of the sun's energy is reaching you? If you're not sure, try this: Hold a flashlight directly above a piece of white paper, as shown.

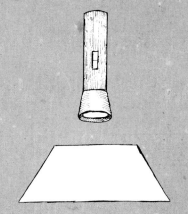

Outline in pencil where the brightest light shines. Now change the angle of the flashlight and repeat. When the light is straight above the paper, all of its energy strikes a small area. The same energy spreads out over a greater area when the light hits at an angle.

Sunlight strikes the Arctic tundra at an angle all year. The sun never rises very high in the sky. And, like a mirror, the white snow reflects back much of the sun's energy. No wonder it is so cold in the tundra. No wonder winter lasts so long. No wonder the snow doesn't melt.

They're back! The arrival of geese, ducks, swans, and other birds is a sure sign winter is almost over.

The male snow goose hisses to warn the caribou not to come any closer. He and his mate are about to build a nest on the part of the small square that is now their territory.

16

On the Move

March. April. May. According to the calendar, it should be spring. Yet 8 inches of snow is still on the 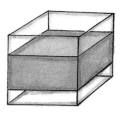 ground! There wasn't one April shower. So far there isn't one May flower. Winter's grip on the tundra remains firm.

But a change is taking place. The days are getting longer and longer. By the middle of May, the sun shines on the square for almost 20 hours a day. It's still too cold for the snow to melt, but much less bitter than it was. The animals can feel the snow softening. Very soon it will start to disappear.

Something else is different. Geese are flying in and landing. These birds aren't lost. They know exactly what they are doing. After spending winter in warmer places such as Florida and Mexico, the geese have flown more than 2,000 miles to reach the tundra.

Day after day they keep arriving. Finally the melt is under way. As soon as a patch of the small square clears, geese fight over it until one pair wins. Now it is their territory, the most important spot on Earth to them. On it

A male snowy owl flying above has caught the eye of this female. He must show off before she will mate.

Just Dropping By?

In spring, look for a flash of yellow or red in a tree in a nearby park. It might be a bird stopping to feed and rest on its long flight to the tundra. This movement of animals from place to place as the seasons change is called migration.

To find out which migrating birds may stop where you live, look in a field guide of birds. In this book are names and pictures of different birds as well as information about where each kind spends winter and summer. You may have a field guide at home. If not, check at the library.

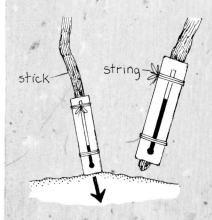

stick

string

If You Were a Lemming

If you live where it snows, tie an outdoor thermometer to a stick and on a very cold day, measure the air temperature. Then stick the thermometer in the snow and measure the temperature near the ground. If you were a lemming, where would you want to be?

they will build a nest, lay eggs, and raise their young.

They won't be alone. By the time the first geese have landed on the tundra, millions of birds are already flying north. Ducks, swans, gulls, sandpipers, hawks, songbirds —all are heading to the tundra to nest.

Not only birds are on the move. Herds of large deer are walking to the tundra. They are caribou, whose close relatives in Europe and Asia are called reindeer.

The caribou left the forest, their winter home. First the adult females started north, then weeks later the adult males and the young. The journey is long and hard, but the caribou are up to it. Thickly furred, the caribou have hooves that spread out to act like snowshoes. The same hooves scrape away snow so the caribou can get at lichens

Some loons and swans build floating nests on tundra lakes.

The male snow goose means business. His job for the next 25 days is to make sure no predator gets near the female, which sits warming the eggs in the nest.

to eat. And like their antlers, the hooves are powerful weapons against wolves and other hungry predators the caribou can't outrun. For the females, predators threaten more than one life. Almost all the females are nearly ready to give birth.

In early June a caribou calf is born in the small square. Within two hours it is up walking on its wobbly feet. In a day it can run faster than you can. Even so, it stays close to its mother. She offers protection and, of course, milk.

Meanwhile, wolves, foxes, and ermines have had babies. Under the snow the lemmings born months ago have had their own babies too. The snow is vanishing. Winter is ending. The sounds of new life fill the small square and the rest of the tundra.

Hidden in her den behind rocks in the square, an Arctic fox feeds her newborns. Soon the babies will be itching to poke around in the chilly world outside.

Stop, thief! Too late. The male fox has stolen an egg out of a bird's nest. He is taking it to the female in the den.

The Race Is On!

At last, all the snow is gone. After nine months, winter has turned into summer. The musk oxen no longer need such heavy coats for warmer days. So their inner fur layer falls out. Everywhere the oxen go, the ground is muddy. Even now, the soil just doesn't dry out in the sun.

If you were in the small square, you could quickly find out why. All you would have to do is dig a hole. The first 12 inches or so of soil would be easy to remove. Then you would hit what felt like solid rock. But it isn't rock. It's permafrost—soil that is permanently frozen. The permafrost stops the melted-snow water from soaking any deeper into the ground.

But wait. Couldn't the melted-snow water in the thawed soil become moisture in the air by evaporating as puddles do after rain? Some of it will, but not much. Warm air can hold a lot of moisture, cool air only a little. And summer on the tundra is still very cool. A week with temperatures above 50°F (10°C) is a heat wave for the small square. Most summer days are chillier than that.

Not much moisture in the air means that not much can come out of the air as rain. Still, there's plenty of melted-snow water in the soil to keep plants and animals

Going Nowhere

Ask an adult to help prepare some gelatin dessert. Half fill a clear plastic container with it. When the gelatin gels, remove it from the refrigerator. On top of the gelatin, spoon a layer of chocolate ice cream, followed by a layer of vanilla. The vanilla stands for the snow that covers the tundra in winter. The chocolate is the upper soil, frozen only in winter. The gelatin is the permafrost. Place the container on a table. Watch what happens when the ice cream melts, but the gelatin doesn't. Where can the melted ice cream go?

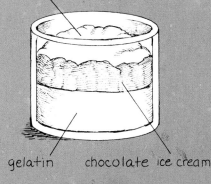

vanilla ice cream

gelatin   chocolate ice cream

Freeze

Fill an empty plastic soda bottle with water and cap it. Place in a freezer overnight. What happens? Does the water inside expand (take up more space) or contract (take up less space)? There is water inside every living cell in every living thing. What do you think would happen if ice formed inside a living cell?

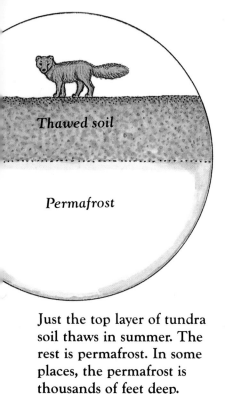

Just the top layer of tundra soil thaws in summer. The rest is permafrost. In some places, the permafrost is thousands of feet deep.

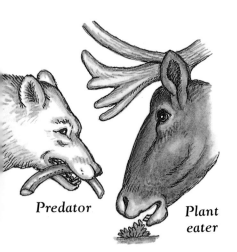

Plant eaters have biting and grinding teeth. In addition to these, predators have long, sharp teeth for tearing into prey.

healthy and to fill the lakes, ponds, and bogs that are all over the tundra.

Day by day, the tundra becomes more and more alive. Patches of plants push up out of the soil. Insects that were frozen are now searching for food. The animals that visited the square in winter are busy "making their living." The foxes, ermines, hares, and lemmings are shedding their winter fur and growing shorter coats for summer. The ptarmigan are replacing their feathers. A quick look at the small square in summer will tell you these animals aren't white anymore. Their new fur or feathers are now perfect camouflage for the tundra without snow.

It is June. There are three months until the cold weather returns. In the small square and the rest of the tundra, there isn't a moment to waste. Eggs must hatch. Babies must grow. Plants must leaf, flowers blossom, and seeds form. Animals that used up their body fat in winter or during the long journey to reach the tundra must eat, eat, eat. All too soon, snowflakes will be falling again. The clock is ticking. The race to beat winter is on.

What's Living on the Tundra?

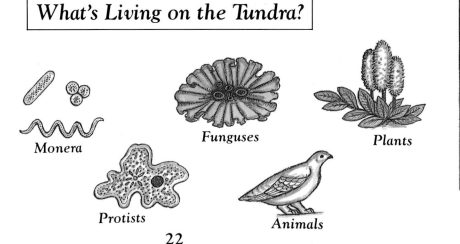

Monera

Protists

Funguses

Animals

Plants

Where you live, most butterflies rest with their wings closed. Tundra butterflies keep their wings open to absorb all the heat they can.

Not only is the ptarmigan perfectly camouflaged, but so are its eggs.

23

This bee knows just where to warm up —in a bowl-shaped poppy. When sunlight bounces off the petals, it focuses on the flower center and on the bee.

Hug That Ground

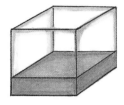

When you go out on a warm spring day, you can look down at plants pushing out of the ground and look up at leaves bursting from buds on tree branches. You may want to watch where you step to keep from squashing a flower. But you needn't worry about crushing a 100-year-old tree.

Not so in the small square—where a tree that old may be just 4 inches tall. It is a dwarf willow with a trunk as wide as a carrot top and branches all too easy to step on. No wonder the tundra seems to be treeless.

None of the plants in the small square reach higher than your knee. By hugging the ground, tundra plants escape the full force of chilly, drying winds. And, as the

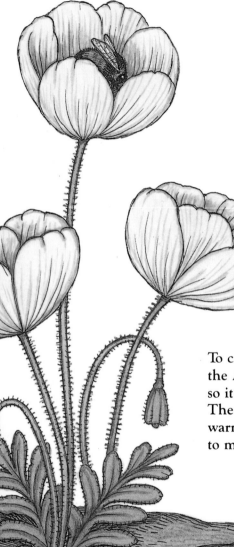

To capture the sun's warmth, the Arctic poppy flower turns so it is always facing the sun. The flower needs as much warmth as it can get, in order to make seeds.

Without a magnifying glass, you might miss springtails, mites, and other small insects—but not a butterfly sunning on a plant.

Dwarf willow

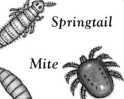

Springtail

Mite

Black fly larva

sun warms the ground, the ground warms the air just above it a few extra degrees—where the plants are growing.

Tundra plants need as much warmth as they can get. Like other plants, they make food by using the energy in sunlight. But like other plants, they cannot make their own heat. The tundra chill slows down food making. Many tundra plants that grow together in thick mats are able to trap any warmth there is. Others are covered in fine "hairs" that do the same thing.

Plants aren't the only ground huggers on the tundra. The insects hatching by the tens of thousands can't make heat either. If they aren't warm enough, their muscles can't work and they cannot fly. But after a bask in the sun on the soil, on a rock, or on a plant, they are soon on the move.

How do lichens help make soil? As they grow, they slowly give off chemicals that wear away rock. That's step one in making soil.

Reindeer moss lichen

Some plants bud one summer, flower the next, and seed a year later. Others bud, flower, and seed each summer.

Looking for Lichens

Lichens grow in the frozen Arctic, in rain forests, along the seashore, and where you live. Search in a park for a colorful patch on a rock, a tree trunk, or the ground. If it is alive, it may be a lichen. Look at the patch under a magnifying glass and draw what you see in your notebook.

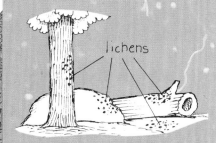

lichens

Sure Signs of Spring

Tundra summers are brief. Where you live, plants may grow year round. Even so, only certain plants push out of the ground when winter ends and spring begins. Draw pictures in your notebook of the first plants of the season. Which hug the ground? Which grow tall? Which show leaves first? Which show flowers first? Try to identify each by using a field guide to plants.

Feb	Mar	April	May

Why come to the chilly tundra for the summer? Lots of food, plenty of nesting space, long days, short nights, and fewer predators than in warmer places.

While a mother ptarmigan nips at tender leaves, her chicks search for tasty caterpillars and spiders to eat.

Now that their egg have hatched, can geese just feed and Not with so many young to protect!

What a Day!

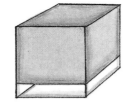

The snow goose on the nest has been more than patient. She felt the first egg crack and the gosling inside slowly peck at its shell, making a big hole in it. She stirred when the other three eggs started hatching soon afterward. She heard her young peeping. But for nearly a day, they have been in no rush to come out of their cramped quarters.

Now, finally, the first gosling is taking that giant step. The others quickly follow. They all look to their parents to take care of them. Their parents know it is time to leave the nest and lead the family to the safety of the pond at the back of the square.

The walk to the water is short but dangerous. While the goslings nip at plants and snap up mosquitoes, the parents are on the alert. They don't want one of their fluffy four to fall prey to a gull suddenly swooping down or to the fox eyeing them from its den.

The ptarmigan nibbling willow buds ignore the geese. So do the caribou grazing on grasses, the Arctic hare munching on a twig, and the lemmings taking bites out of just about every plant they find.

Throughout the day, plant eaters visit the square to snack. Then they move on in search of more food. Because the animals do not overeat any one patch of tundra, the plants get a chance to keep growing and stay healthy.

The fox can try to catch a baby bird. But it can't do a thing about biting mosquitoes and flies that won't leave it alone.

Mosquito

Black fly

This lemming should have been more alert while feeding. An ermine caught it. But the ermine doesn't notice the owl about to grab it. The owl gets dinner; the lemming escapes.

Two young caribou run from pack of wolves circling an old weak male. The prey may try fight, but it won't win.

The hours pass: 8...9...10...11 o'clock. Midnight. The sun hasn't set. It is nighttime, isn't it? Not anymore! From about the middle of June through the beginning of July, no night will come to the small square. There will be one very long day.

No animals seem to miss the darkness. Round the clock, they can find plenty of food to fill their stomachs and to feed their fast-growing young. There are more than enough lemmings for hungry foxes, wolves, and snowy owls. There are also lots of hares, musk oxen, caribou, and ptarmigan for predators to eat.

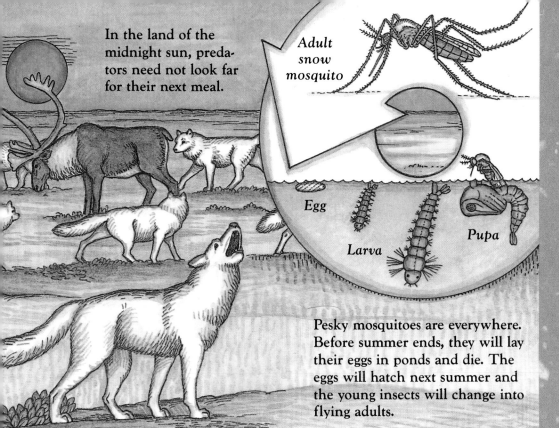

In the land of the midnight sun, predators need not look far for their next meal.

Adult snow mosquito

Egg

Larva

Pupa

Pesky mosquitoes are everywhere. Before summer ends, they will lay their eggs in ponds and die. The eggs will hatch next summer and the young insects will change into flying adults.

Playful fox pups frolic close to their den while their parents are off hunting. By chasing and pouncing on each other, the pups are already learning skills they will need when they must catch prey on their own.

All is not fun and games on the tundra. Billions of mosquitoes, black flies, and other biting insects attack both furry mammals and birds. It's bad enough that these insects suck blood when they bite. But the itch, swelling, and pain from so many bites never seem to go away.

By the time the sun starts to set again, life in the small square is at its fullest. But when the very long day is over, winter is much closer.

Let's Tilt Again

Repeat the activity on page 13. Focus on the orange when it is tilted toward the light. When the top half of Earth is tilted toward the sun, there is no night, just day, in the Arctic tundra. As Earth keeps moving around the sun, night returns to more and more tundra. But for half the year the North Pole has day without night.

Lemming Math

How quickly do lemmings multiply? A pair can have 10 babies at one time. A month later the babies are ready to mate. If all the babies find mates, there may be 10 times 10, or 100 new babies by the next month. How many will there be if those 100 new babies have 10 babies each? And don't forget that the original pair may have more babies. What do you think keeps the lemmings from taking over the world?

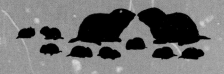

The red on the male caribou's antlers is a thick, soft skin called velvet. Once a year caribou lose and regrow their antlers. The velvet protects growing antlers, then peels off.

My, how those baby snow geese have grown! Soon they'll be flying toward their winter home, which they've never seen.

30

On the Way Out

The days are shorter. The winds are brisker. The air is so cold that there is often frost on the ground in the morning. August is over on the tundra.

A contest is about to begin in the small square. Two male caribou are facing each other and roaring loudly. They lower their heads and lock antlers. With all their strength they push, shove, twist, and turn. Each one tries to force the other to back off and leave. The struggle is exhausting. When the weaker male senses he is losing his balance, he gives up rather than injure himself. The stronger caribou has won the square as part of his territory. He alone will mate with the females that enter it.

Like the caribou, the birds are restless—but not to start new families. Instead, the birds are busy eating as much as they can. Soon most will head south for the winter. The extra body fat they have put on will be their fuel for flying thousands of miles.

You can guess which animals are turning white again and growing thicker fur or feathers. The body fat they add will help them keep warm when they need warmth most. As for the grizzly bear and ground squirrel, they must fatten up enough to last through nine months of sleep.

Something else is happening. The plants are changing color. Summer is definitely on the way out.

Colorful berries contain seeds ripening inside them.

A Matter of Fat

How does the body fat that tundra animals add in summer—and isn't used for energy—help them stay warm in winter? To find out, make a fat glove. Spoon a pound of solid vegetable shortening (such as Crisco) into a 1-gallon plastic bag. Fill a bucket with cold water. Now place your hand inside an *empty* 1-gallon plastic bag. Seal the bag around your wrist with masking tape. Dip your hand into the water. How does it feel? Pull your hand out and dry off the bag. Put your bagged hand into the vegetable shortening and seal the second bag shut around your wrist with more tape. Dip your double-bagged hand back into the water. How does it feel? The shortening works the same way as a layer of fat under an animal's skin. It should—shortening is made of fat.

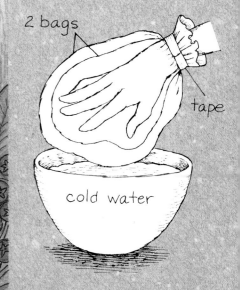

2 bags

tape

cold water

Ready or Not

The small square is ablaze with color—deep reds and oranges and rich golds. Autumn on the tundra is one of the briefest on Earth. It lasts only a few weeks in September. And if there is an early snow, it doesn't last even that long.

The snow geese that hatched in the square all made it through the summer without being eaten. With their parents, they have taken to the skies. So have the millions of other migrating birds. On the way to their winter homes, some may land where you live. Look for them. A field guide will help you identify these visitors.

Up, up, and away. Off go the geese and swans with their families. This flight is very important for the young. Their parents will show them the exact route to their winter home.

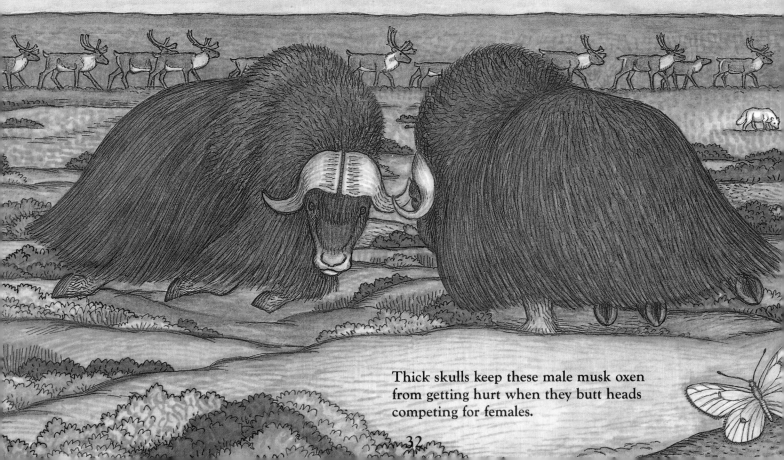

Thick skulls keep these male musk oxen from getting hurt when they butt heads competing for females.

With each passing day, the caribou gather in herds. They will drift back to the forest, trailed by hungry packs of wolves hunting for a big meal.

The square, however, is far from deserted. Foxes, bears, and ptarmigan feast on seeds and berries that have ripened with the change of season. Male musk oxen crash head-on, fighting over females. The ground squirrel digs out a burrow in which to sleep away the winter.

Many caterpillars and other insects are also digging into the soil. Their bodies are already making natural "antifreezes," chemicals that will keep them from dying when they are frozen for nine months. Meanwhile, bacteria, protists, and funguses in the soil help break

Most of the time ptarmigan walk instead of flying. When scared by an ermine, the birds take to the air.

Some of the sun's energy is stored in food plants make. In winter that energy keeps plants alive. And plant eaters. And predators that eat them.

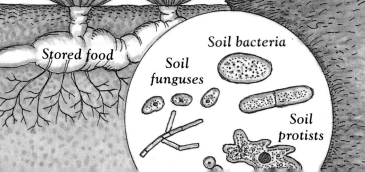

Stored food

Soil bacteria

Soil funguses

Soil protists

The ground squirrel digs its burrow where the ground rises and the soil is driest. First it digs down, then turns and tunnels up. At the end of the tunnel, it builds a nest of grasses. The tunnel and the grasses help trap the little heat the squirrel's body gives off as it sleeps through winter.

Polar bear

Grizzly bear

As summer ends, both a polar and a grizzly bear might visit the small square. In the dead of winter neither is to be seen.

Twelve thousand years ago, woolly mammoth may have tromped across the small square. Don't look for one today. These huge mammals are extinct.

down dead leaves and animal droppings. They recycle nutrients that plants will need next year. Once the soil is frozen solid, this rotting process will just about stop.

Plants, too, are preparing for the harsh months ahead. They store food underground in their roots and bulbs. Some plants are growing buds that will open the moment the snow melts, ushering in next summer.

As the air turns colder and colder, the less moisture it can hold. The air has so little moisture that only about 10 inches of snow will fall during the entire winter. That won't be enough for the female polar bear walking across the square. She has come from the coast in search of deep snow. She needs to dig a den where her cubs will be born. Because the snowfall in the small square won't be enough for her, she moves on.

The small square is quiet. The first snowflakes are falling. Ready or not, winter is back.

Time to Explore

The small square seems deserted again. The lemmings and voles have disappeared under the snow. Who can blame them for seeking the warmest place they can find? If you were there, you too would do whatever you could to escape the deadly cold.

Small Square in Summer

Can you match each living thing with its outline?

Caribou

Snow bunting

Brant geese

Willow ptarmiga

Rock ptarmigan

Dwarf birch

Snowy owl

Lichens

Forget-me-not

Whistling swan

Sedge

Ermine

Swarm of
black flies

Arctic
fritillary

Arctic ground
squirrel

Tussock
grass

Snowshoe
hare

Cotton grass

Arctic fox

Red phalarope

Tundra vole

Arctic poppy

Collared lemming

Dwarf willow

Bumblebee

Snow geese

35

Tundra Diorama

Take a shoebox and measure its length and height. Cut a piece of paper for the background wall about ¼ inch shorter than the box height and about 4 inches longer than its length. On it you can draw a pack of wolves, a herd of musk oxen, an aurora for a winter night, or a herd of caribou and migrating birds for a summer day. Place the picture in the box and tape each side to the front. The picture will curve.

On separate sheets draw and color tundra plants and animals, each with a flap at the bottom. If you are making a winter scene, draw the animals on snow. Cut each picture out, bend its flap, and glue or tape it to the bottom of the box. You can also draw and color a rock with a fox den inside—or use a real rock.

Summer is different. It is a great time to explore the far north. Now that you have solved so many tundra riddles, you are ready. You know why there is no need to pack a bathing suit, shorts, or an umbrella. You will be sure to take along a jacket, a sweater, a hat, and gloves. And boots, of course, to tromp through mud. You also know to be on the lookout for anything that will keep the mosquitoes and other bugs from taking bites out of you. Bring your tundra notebook along. Don't be surprised if you don't find all the plants or animals from this book in your small square, or if you find something not in this book.

If you do visit the tundra and think there are an awful lot of lemmings, you may be right. As long as there are plenty of plants to eat and few predators, lemmings keep having babies and more babies.

Too many lemmings, though, can be harmful. They nibble down plants until there isn't enough food for them all. Many die. Others leave the tundra in search of more plants and never return. The next year there are hardly any lemmings for predators to eat. The predators can't find food, have few or no babies, and flee the tundra or die. The plants recover and, with fewer predators around, the number of lemmings rises again.

Tundra plants and animals cannot afford to lose their food supply or have their homes ruined. So enjoy your small square when you visit, but leave it the way you found it. Nature will do the rest.

> Small Square in Winter

Can you match each living thing with its outline?

36

Arctic willow

Willow ptarm

Rock ptarmig

Reindeer moss lichen

Raven

Musk ox

Gyrfalcon

Arctic wolf

Snowy owl

Ermines

Lichens

Arctic ground squirrel

Collared lemmings

Redpolls

Arctic fox

Showshoe hare

Tundra vole

Arctic shrew

See You in Siberia

On some small squares of Siberian tundra, Siberian white cranes nest. But not many. These birds are an endangered species.

Not all small squares of tundra are the same. The square explored in this book might be found in the north-central part of Canada. Similar squares occur in north-western Canada and northern Alaska.

Across the ocean in Siberia (part of Russia), there is tundra too. It is home to many plants and animals you already know about: lemmings, Arctic foxes, snowshoe hares, reindeer, musk oxen, and dwarf willows. But don't let that stop you from wanting to visit. Where else could you come across the Siberian white crane and the Siberian salamander?

Ross's gulls hunt small fishes in ponds to feed their newly hatched chicks.

Small seed eaters from Europe and Africa flock to Russia for the tundra summer.

About 4,600 years ago, an artist painted a flock of birds on a tomb in Egypt. Those birds were red-breasted geese, which still nest in the Siberian tundra.

The only amphibian that spends the winter frozen is the Siberian salamander. Unlike mammals and birds, amphibians cannot make their own body heat.

All these animals are vertebrates—they have bones in their bodies.

During the tundra winter, mammals and birds can stay active because they make their own body heat. They are the only kinds of animals that do so. Fur on mammals and feathers on birds help keep body heat from escaping.

Musk ox

Mammals

Arctic fox

Arctic ground squirrel

Arctic shrew

Ermine

Collared lemming

Grizzly bear

Arctic wolf

Polar bear

Tundra vole

Caribou

Snowshoe hare

Raven

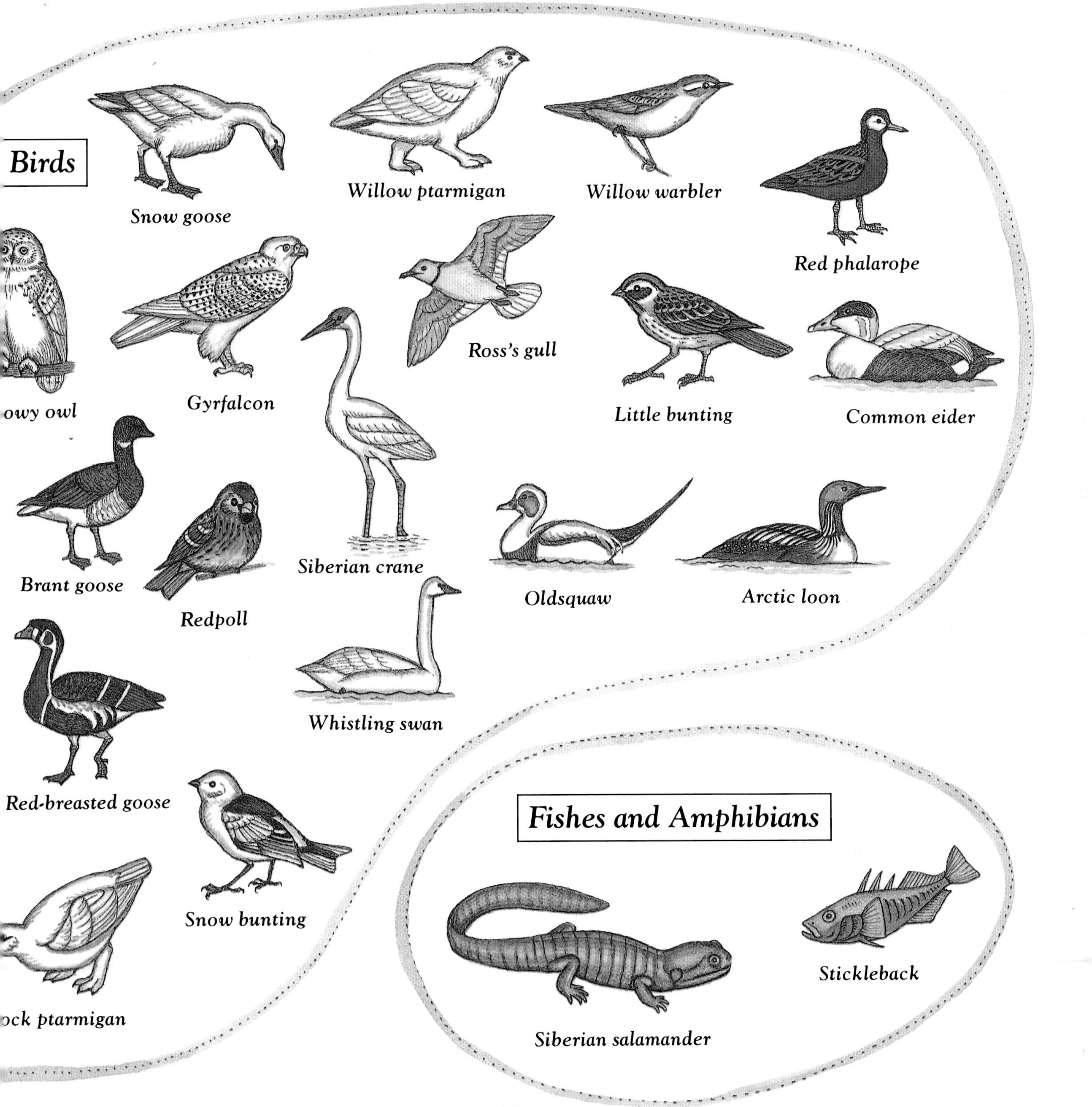

Birds

Snow goose

Willow ptarmigan

Willow warbler

Red phalarope

owy owl

Gyrfalcon

Ross's gull

Little bunting

Common eider

Brant goose

Redpoll

Siberian crane

Oldsquaw

Arctic loon

Red-breasted goose

Whistling swan

Snow bunting

ock ptarmigan

Fishes and Amphibians

Siberian salamander

Stickleback

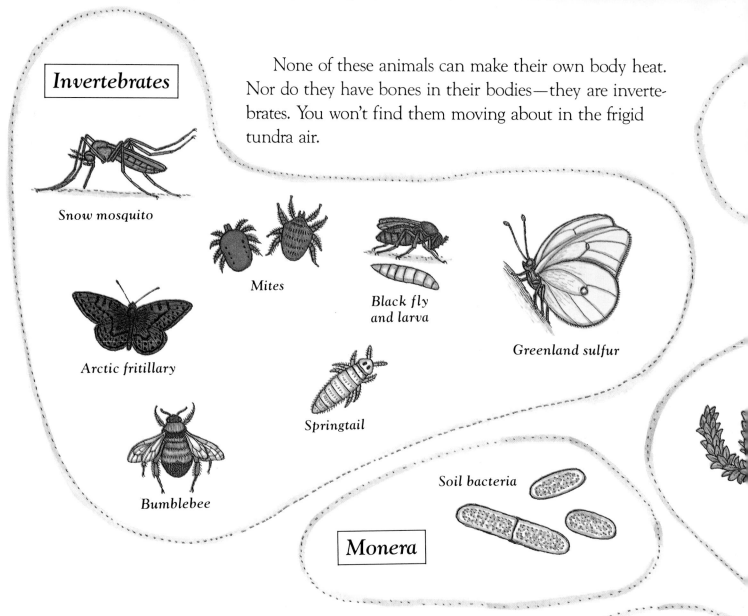

Invertebrates

None of these animals can make their own body heat. Nor do they have bones in their bodies—they are invertebrates. You won't find them moving about in the frigid tundra air.

Snow mosquito

Mites

Black fly and larva

Greenland sulfur

Arctic fritillary

Springtail

Bumblebee

Monera

Soil bacteria

Plants make food by using energy from the sun. But funguses, such as those in lichens, cannot. You can examine plants and funguses with a magnifying glass. To see one-celled monera and protists, you need a microscope.

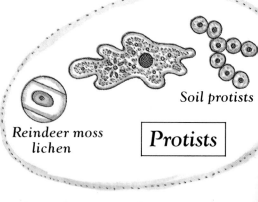

Soil protists

Reindeer moss lichen

Protists

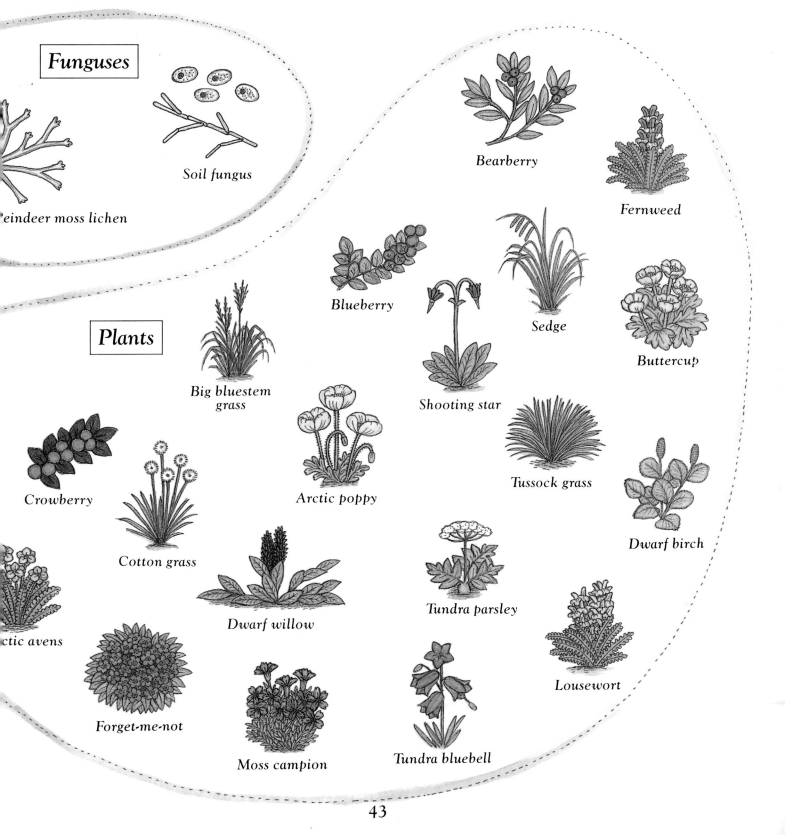

Funguses

Reindeer moss lichen

Soil fungus

Plants

Bearberry

Fernweed

Blueberry

Sedge

Buttercup

Shooting star

Big bluestem grass

Arctic poppy

Tussock grass

Crowberry

Cotton grass

Dwarf birch

Dwarf willow

Tundra parsley

Arctic avens

Lousewort

Forget-me-not

Moss campion

Tundra bluebell

Index

A
adaptation 5. *Any part of a living thing that makes it fitted to survive where it lives.*

Alaska 38

alga (AL-guh) *plural* **algae** (AL-jee) 7. *A kind of protist that can make its own food. Also, a kind of plant.*

amphibian (am-FIB-ee-in) 39, 41. *Bony animal that lives the first part of its life in water and the second part on land.*

antifreeze 33

antler 19, 30, 31

Velvet

Arctic Ocean 6

aurora borealis (uh-RAWR-uh bawr-ee-AL-iss) 13, 36. *"Curtains" of colorful light hanging in the sky, mostly seen in the far north.*

autumn 8, 32

B
bacteria 33. *Kinds of monera —one-celled creatures that don't have a nucleus (control center).*

beak 8

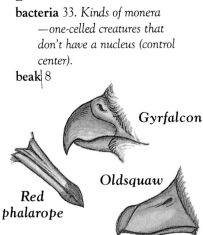

Gyrfalcon

Oldsquaw

Red phalarope

bear 3, 6, 10, 14, 31, 33, 34

bee 24

berry 8, 31, 33

bird 5, 8, 10, 16, 17, 18, 27, 31, 33, 36, 39, 40, 41

bog 22. *A kind of land that is naturally wet.*

branch 24

bud 8, 24, 25, 27, 34. *The tip of a plant, from which leaves, flowers, or stems grow. (Also, to grow buds.)*

Dwarf willow

Dwarf birch

bulb 34. *An underground bud, surrounded by thick leaves, that stores food for some kinds of plants.*

(continued)
bull 11. *Name given to the adult male of the musk ox (and some other mammals, such as elephants, cattle, and moose).*

burrow 33. *A hole or tunnel an animal digs in the ground. (Also, to dig that hole or tunnel.)*

butterfly 23, 24

C
camouflage (KAM-uh-flahj) 8, 22, 23. *An animal's color, pattern, or shape that helps it hide in its surroundings from other animals that want to eat it.*

Canada 38

caribou (KAR-uh-boo) 6, 16, 18, 19, 27, 28, 30, 31, 33, 36

caterpillar 7, 15, 26, 33

cell 21. *Smallest living part of all plants, animals, and funguses. Some living things are made up of just one cell.*

claw 8

cow 11. *Name given to the adult female of the musk ox (and some other mammals, such as elephants, cattle, and moose).*

crane 38

D
deer 18

den 10, 19, 28, 29, 34, 36. *A cave or other shelter where an animal lives, sleeps, or hides.*

down 10

droppings 34

duck 16, 18

Index

E
Earth 13, 17, 29
egg 18, 19, 22, 23, 26, 27, 29
endangered species 38
energy 9, 10, 15, 25, 31, 33, 42. *Ability to do work or to cause changes in matter.*
ermine (UR-min) 15, 19, 22, 28, 33
evaporation 21. *The process of changing from a liquid into a gas.*
extinction 34

F
falcon 9
fat 10, 15, 22, 31
feather 9, 10, 12, 22, 31, 40
field guide 17, 25, 32
fish 38, 41
flower 5, 17, 22, 24, 25
fly 27, 29
food 7, 9, 10, 12, 14, 25, 26, 27, 28, 33, 34, 36, 42
foot 8, 9, 10, 19

Musk ox

Snow goose

Arctic wolf

forest 6, 18, 33
fox 3, 6, 9, 12, 14, 15, 19, 22, 27, 28, 29, 33, 36, 38

Arctic fox

Red fox

fritillary (FRIT-uh-lehr-ee) 24
frost 31
fungus 7, 22, 33, 42, 43
fur 9, 10, 11, 12, 18, 21, 22, 29, 31, 40

G
gas 13
goose 16, 17, 18, 26, 27, 30, 32, 39
guard hair 11
gull 18, 27, 38
gyrfalcon (JUR-fal-kin) 8

H
hare 3, 8, 9, 22, 27, 28, 38
hatching 27, 29
hawk 18

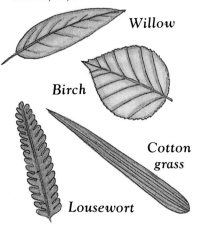

Rough-legged hawk

heat 9, 10, 11, 23, 25, 33, 39, 40, 42
herd 33, 36
hibernation 15
hoof 18, 19
horn 10

I
ice 3, 21
insect 14, 22, 24, 25, 29, 33
invertebrate 42

L
lake 18, 22
larva (LAHR-vuh) 29. *Very young insect or other animal, such as a frog, that changes a lot before it looks like its parents.*
leaf 22, 24, 34

Willow

Birch

Cotton grass

Lousewort

lemming 3, 12, 14, 15, 17, 19, 22, 27, 28, 29, 34, 36, 38
lichen (LY-kin) 7, 18, 25, 42
loon 18

M
magnifying glass 5, 24, 25, 42
mammal 29, 39, 40

Index

mammoth 34

microscope 42. *An instrument through which people can see many things that they can't with just their eyes.*

midnight sun 29

migration 17, 32, 36

milk 14, 19

mineral 7

mite 24

moisture 21, 34. *Water in the air —such as tiny droplets that make up the clouds and water vapor, an invisible gas.*

monera (muh-NEER-uh) 22, 42. *Creatures made up of one cell that doesn't have a nucleus (control center).*

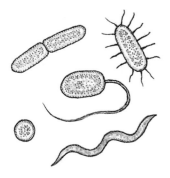

moon 12, 13

mosquito 27, 29, 36

moth 7

mud 5, 21, 36

musk ox 8, 10, 11, 12, 21, 28, 32, 33, 36, 38

N

nest 16, 18, 26, 27, 38

Whistling swan in nest

Brant goose in nest

Tundra vole in nest

Snowy owl in nest

night 10, 12, 26, 28, 29

North Pole 13, 29

notebook 9, 25, 36

nutrient (NOO-tree-int) 34. *Any part of food that living things must have to build cells or to use as a source of energy.*

O

owl 12, 17, 28

P

pack 12, 13, 33, 36

paw 12

permafrost 21, 22

petal 24

plant 5, 10, 14, 15, 22, 24, 25, 27, 31, 33, 34, 36, 38, 42

pond 22, 27, 29, 38

poppy 24

predator (PRED-uh-tur) 8, 9, 15, 18, 19, 22, 26, 28, 29, 33, 36. *Animal that kills other animals for food.*

prey 14, 22, 28, 29. *Animal hunted or caught for food by a predator.*

protist 22, 33, 42. *Creature usually of one cell that has a nucleus (control center).*

Nucleus

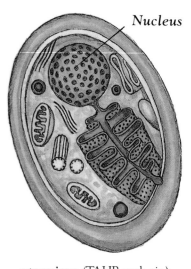

ptarmigan (TAHR-muh-gin) 8, 9, 22, 23, 26, 27, 28, 33

pup 29

pupa (PYOO-puh) 29. *Part of the life cycle of some insects in which their bodies change completely as they become adults.*

Q

qiviut (KEE-vee-it) 11. *Eskimo word for a musk ox's soft, woolly inner layer of fur.*

Index

Baird's sandpiper

Wolf skull

Arctic fox

Caribou

Tundra vole

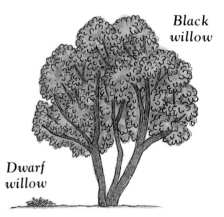

Black willow

Dwarf willow

The blue area on this map shows where tundra exists around the world. The small square in this book might be found in the green area.

Further Reading

To find out more about migrating birds, look for the following in a library or bookstore:

Golden Guides, Golden Press, New York, NY

Golden Field Guides, Golden Press, New York, NY

The Audubon Society Beginner Guides, Random House, New York, NY

The Audubon Society Field Guides, Alfred A. Knopf, New York, NY

The Peterson Field Guides, Houghton Mifflin Co., Boston, MA

Reader's Digest North American Wildlife, Reader's Digest, Pleasantville, NY

937